Renewable Energy

You Can Make a Difference Too!

by

Owen Jones

Renewable Energy

Copyright

Published by Megan Publishing Services
https://meganthemisconception.com

Copyright Owen Jones 2024 ©

Owen Jones

Hello and thank you for buying this ebook called '**Renewable Energy - You Can Make a Difference Too!**'.

In the face of climate change and the urgent need to transition towards sustainable practices, the rôle of renewable energy has never been more critical. The good news is, you have the power to be a part of this transformative journey. Welcome to "Renewable Energy - You Can Make a Difference Too!" - a guide designed to empower you with practical knowledge and actionable steps towards a greener, more sustainable future.

Within these pages, we'll explore the diverse world of renewable energy sources, from solar and wind power to hydropower and geothermal energy. We'll unravel the benefits they offer, not only for our environment but also for your household or business. You'll learn how embracing renewable energy can lead to reduced energy bills, increased energy independence, and a lowered carbon footprint.

But it doesn't stop there. We'll delve into the practical steps you can take to integrate renewable energy solutions into your daily life. Whether you're a homeowner, a business owner, or simply someone passionate about the planet, this booklet will provide you with clear, actionable advice on how to make sustainable choices that align with your goals and values.

The journey towards a greener future begins with you. Together, we can harness the power of renewable energy to create a cleaner, more sustainable world for generations to come. So, let's embark on this empowering journey, and discover how you too can make a significant difference through renewable energy!

Regards,
Owen Jones

Owen Jones

Table Of Contents

1. Converting Your Home Into A Solar Home

When it comes to converting your home into a solar home, there are several options, because not all homes have the same problems, the same requirements or the same potential sustainable power sources. Therefore, if you are going to try a total conversion or even get off the grid entirely, you will either need to do some research or call in an expert to make a survey for you.

If you call in an expert, try to get an independent one, so that you can work out the costs of fulfilling your energy needs yourself. You will have to pay for such a survey, naturally, but you could carry out a report yourself with a bit of work on your behalf. In order to produce a solar home, you may find the rest of this article interesting.

There are fundamentally two types of solar design:

passive and active solar energy. Passive solar energy can be used to supply heating, cooling and natural light for your home. Active solar energy is used for powering home appliances, tools and lighting. It is the ideal combination of these two types of solar energy that you will attempt to accomplish, if you are attempting to convert your home into a solar home.

You can use passive solar energy methods in many ways, although they are more easily built in during the actual construction phase of a new home. The largest area of glass should face south or be within 30 degrees of due south. This will catch the maximum amount of heat. This heat can then be circulated around the house by stone floors and stone walls.

The central heating ducting and furnace fan can be used to assist, if necessary. If the house becomes too hot in the summer, awnings or even solar panels could be dropped down in order to put the windows in shadow. When thinking of passive solar energy, you should try to think of means of supplying warmth and coolness without using electricity. For example, a skylight at the top of the stairs will permit the warmest air in the house to escape, since hot air rises. This will cause cooler air to be sucked into the house at lower levels.

The other feature of a solar house is the generation of electricity by the use of solar panels grouped into solar arrays. Solar panels make use of photovoltaic cells to convert light into energy. This energy can then be used to power everyday electrical appliances of all sorts or some or all of it can be stored in batteries for later use. Conversion from AC (alternating current) to DC (direct current) and back again, if needed, is a straightforward affair.

Solar energy can also be used to warm up water for the pool or for the home. The most usual sort of system uses pipes filled with a kind of anti-freeze to collect the sun's heat and pass it on to tanks of water by means of a heat exchanger.

A solar home uses energy efficiency to reduce the need for heating, cooling and electricity. The use of higher grade insulation, more energy efficient windows, kitchen appliances and lighting than those used in traditional homes, will save you a lot of money and energy.

As you can see, some of these modifications, particularly the passive ones are structural, so hard to apply in many homes, but there is always something

Renewable Energy

you can do to cut your energy bills and slowly convert your home into a solar home.

2. How To Reduce Our Oil Addiction

The world has been on a collision course with disaster with reference to its oil dependence since at least the Second World War and that finished almost seventy years ago! As if that wasn't enough to get through the skulls of our 'leaders', we were reminded with the Suez Crisis in the Sixties, the Oil Crisis of the Seventies and the problems with Saddam Hussein and Iraq in the Nineties and Noughties.

And we still didn't learn. Our successive governments of every colour just turned a blind eye to our oil consumption and allowed their pals in the fossil fuel industries to flog us the stuff that they had mined or drilled from the planet. Why didn't the West plan full-scale research into alternative technologies like geothermal, wind and solar power?

The answer can only be greed and personal, self interest. Surely they knew that fossil fuels would never become cheaper? What commodity does when it is in ever diminishing supply? These guys are experts in capitalism - they knew that, with decreasing supply and increasing economic demand from three billion Chinese and Indian people, the price of energy could only go up. And not slowly either.

Instead of being more or less self-sufficient in our energy needs, after seventy years of warnings, we in the West are just as reliant on dangerous or unstable countries in the Middle East as ever. Perhaps even more so than before, although some governments have started to look for the life jacket now that water is actually seeping under the cabin door.

They are still only hoping that the ship won't go down though. They cannot guarantee it with any degree of certainty. If Russia turned off the gas and the petroleum dried up, France might have to stop selling nuclear generated electricity and then where would Europe be? America, while not relying on Russia and France for natural gas and electricity, would not be far behind in the queue to panic buy fossil fuels.

We need positive action from people with vision and

determination. No politician from the Twentieth Century has proved that he or she has it. Don't elect any of them any more. Try younger, hungrier, less wealthy candidates. What on Earth are we doing electing oil barons to fix a fuel crisis that they have presided over for a hundred years? And why are we electing members of rich banking families to fix a problem that the created?

The logic of the voting population in the United States and Europe is unfathomable. Except to politicians. They know that we have behaved stupidly in the past and sucked up all their lies and excuses. Let's prove them wrong next time and send the buggers packing!

We, the citizens of the world do not need lying, thieving, corrupt and incompetent self-seekers to ruin or world for us. We can do that ourselves if we wanted to. Let's give a new generation of young politicians a chance to run our economy, but keep a close eye on them.

Don't let's allow them to get away with anything like we did the last lot. We the voters took our eyes off the ball and look what we've got for our complacency a ruined environment and a ruined economy!

3. Nuclear Energy - Is It A Green Solution?

The West consumes far more than its fair share of fossil fuels. America has the highest per capita rate of consumption and it has to stop or rather be reduced. As a rich country, America is buying up rare resources from poor countries just because it has the money to be able to do so.

This sort of greedy behaviour by Western governments on behalf of its people is not only unfair, it is also immoral. It is tantamount to a rich family buying up black market goods during rationing while the rest of the population is struggling to do without. It is allowing the rich to do what they want and not caring about the future of the poorer countries.

This is bad enough, but the Western governments' lack of a sensible energy policy is ruining the global environment. The greenhouse gases that are created ad nauseam in the West know no boundaries. Pollution can be airborne or travel by sea. We are polluting countries that are doing no or little harm to their own environment. How can that be fair?

Excessive greenhouse gases has the effect of warming the global climate in the air and in the sea. Those of us who live in cold countries may think that climate change is marvellous, but what about those people who live in hot countries? Their countries could become so hot or so dry through global warming that they can no longer produce enough food to feed themselves. That cannot be fair.

Smog and pollution go hand in hand with excessive greenhouse gases and they produce illness, especially respiratory diseases and allergies, but again, not only in the country that has produced the pollution. The excessive use of fossil fuels puts their price up too. Rich countries don't like to have to pay more, but they can. /Poor countries must pay the same price, but cannot afford it.

Another problem that comes from being over-reliant on traditional fossil fuels is that none of the Western countries is self-sufficient in them. This means that

Western industrial powers are reliant on and could be held to ransom by the net sellers of those fuels. The countries that sell fuel tend to be undeveloped, dictatorial, unstable countries. So in effect, it is these countries that call the shots in our countries and will continue to do so until we are self-reliant.

Many countries have turned to 'traditional' alternative energy sources like wind, solar and geothermal power, because nuclear power plants got a very bad press what with problems in the UK, Russia, America and Japan over the decades. However, the generation of nuclear power may be the only way forward when you consider the huge amounts of electricity that we need to continue and improve our lifestyles.

France and Finland among others have had very successful nuclear power plant strategies, so it seems that the time has come to take another look at using nuclear reactor stations to feed the national grid, while using solar, wind and geothermal power production locally.

Renewable Energy

4. Alternative Energy Overview

It seems like significant efforts to find the best alternative energy sources are being made by lots of countries as well as the USA and even many of America's cities. One proof is the signing of the Kyoto Treaty, the foremost goal of which is to reduce greenhouse gases and pollutants.

Sustainable energy sources have proven to be of great help in reducing the amount of poisons, which are largely by-products of the use of fossil fuels. Sustainable sources also conserve the natural resources that people use as sources of energy. What are the most prevalent streams of renewable energy? Here's a list to give you a basic knowledge of the topic.

1. Solar Power. This works by using the sun's rays to charge solar power batteries. The process converts the light coming from the sun into electricity. When

the sun's rays hit solar thermal panels, the power is then used to heat air or water. The sun's rays can also hit parabolic mirrors. This process can produce steam by heating water. But you don't need all these scientific processes to be able to benefit from solar power. All you have to do is open the windows and blinds of your room to let the sunshine in. You'll get an instantaneous heater without having to use any means for the conversion of this energy supply.

To date, the main disadvantage of using solar energy was that it is restricted. You cannot use it at night time or during days when it is raining or even cloudy. This has been partly improved through the use of solar power stations. But they are so expensive that there aren't many about.

2. Wind Power. The energy in the wind turns the blades of wind turbines to produce electricity. Electricity is produced through the use of an electrical generator. In the old days, windmills were used so that machinery could replace human physical labour This included the pumping of water and the milling of grain which were important to farming.

Now there are large scale wind farms that produce electricity. The end product is then dispersed through the national grids and small systems owned by private individuals to distribute electricity to far-flung areas and homes. There are many benefits to this kind of power. The major one, of course, is that it doesn't create any by-products detrimental to the environment. And we will never run out of this source of renewable energy.

3. Geothermal Energy. This comes from underground. Holes are drilled in key locations and the hot rocks beneath produce steam. This is then cleansed to be used to drive turbines. The latter then become a power source for electric generators.

To make sure that no harmful by-products are created in the procedure, geothermal plants must be designed carefully, but once they have been set up, they are self-sufficient in creating energy.

4. Hydroelectric Energy. This power is created through the use of dams that retain water to drive generators and water turbines. Tidal power can also be used if a dam is not suitable. The idea here is to make use of the kinetic energy of water.

If you read through the available alternative energy sources, you will be amazed at how nature can work miracles. It is your duty to take care of everything around you.

5. Simple Ways To Save Energy At Home

As we all know to our cost, quite literally, fuel bills for our homes go up every year. However, have you ever actually stopped to examine exactly why that is? Sure, energy prices have gone up again, but is that the whole story? There could be other factors at play, couldn't there?

As a house gets older it is harder to keep at the right temperature and likewise, as an appliance gets older, it too struggles to be as efficient as it was when new with the same amount of power. Not only that but newer appliances are made to be more efficient too.

So take a look at the list of simple ways to save energy at home below and see if there is anything that you can do to bring your household energy bills down a bit and save electricity and money at the same time.

The first things to check are your windows and doors. Do they still fit properly? It may sound like a daft question, but houses can move a little with time, especially if they are new or there has been a drought or flood recently. Timber windows and doors can shrink quite a lot in very dry weather and so can their frames causing the heating or cooling in your home to be affected.

The door or window can come away from the frame, but the frame can come away from the wall too. I noticed it in my last house only when I redecorated one year. I stripped the wallpaper and could see outside under the windows in my living room. The only thing between me and the snow outside was a piece of paper!

The proper way to fix this is with cement and then plaster or fine filler, but a decent alternative, if more costly, is to fill the void with mastic or silicone from a tube. All of this stuff can be bought at a builders' merchant or a hardware store. This will kill those indefinable draughts and stop insects coming in for refuge at the same time.

When your doors and windows all fit snugly in the holes that were made for them, keep doors closed if you don't need them to be open, otherwise you are heating or cooling a room that you are not occupying. Likewise, when it gets dark outside, draw the curtains to keep the heat in or out. Heavy curtains are the most efficient for this insulation work.

Do you have a ceiling fan? Do you alter the blades every season? Most ceiling fans have reversible blades or a reversible motor to make them more efficient. In the winter, you want to suck heated air down onto you and in the summer, you want to blow hot air up and away. Don't forget, hot air rises.

While you have that job in your mind, why not reverse the mattress too? Many have a close-knit side for the winter and a looser one for the summer. Likewise, if you have a tiled floor, you could lift the rugs in the summer to get the benefit of their cooling effect and relay them in the winter for insulation.

And last but not least in our list of simple ways to save energy at home, have your heating and cooling systems cleaned and overhauled every year in the season when they are not required, if you leave it until the last minute, you will pay through the nose, like most people, who discover on the first day that they need it that the appliance is not working (well).

6. Declare Independence By Building Your Own Solar Panels

How would you like to get off the electricity grid? What a dream, eh? Sadly, it will remain a dream for many people because the cost of having solar panels installed professionally hardly warrants the outlay. Numerous experts calculate that it can take well over 10 years to recover the cost of the professional installation of solar panels. This is way beyond the horizons of most home owners, who would require a break-even point of three to five years. This is not going to happen in the near future, even with the rapidly rising cost of electricity.

However, there is an alternative. Everybody knows that the labour element in any professional job is equivalent to or even exceeds the cost of the materials in that job, so you could save half the cost just by installing the solar panels yourself. How about saving another 50% or more on the price of the solar panels by assembling them yourself too?

Now we are getting into the realm where the cost of a viable solar power system to take the place of state electricity is credible. It could even pay for itself in a couple of years by reducing your reliance on the grid or even allowing you to come off it altogether. Did you know that the grid will buy your surplus electricity from you too?

This may sound like fantasy, but it is not that hard if you have plans or and a solar panel kit. In fact, the parts needed to make your own solar panels are pretty common these days and you will be able to pick them up either at your local hardware store or at a hobbyists like Radio Shack. If you think that that is too time-consuming, you could just buy a kit. These kits are so uncomplicated that any teenager should be able to assemble one.

Later on, after completing a kit or two, you might have the confidence to buy the parts separately, which will save you even more money. One of these kits would be enough to power a few tools in your shed or the lights in your garage or a pump on the pond or pool. If you grouped a number of them together, you could start to reduce your home's reliance on the grid just by harnessing the energy of the sun. Wouldn't that be wonderful?

Solar panels have been about for a long time and so people remember when they needed strong sun light to be of any use, but public awareness has not kept up with the rate of technological improvement. Solar panels are much more sensitive now and they can produce electricity from light using powerful photo voltaic cells (PV's) - they do not require blazing sunshine to work any more.

So, if you want to go down the road to independence from the electricity grid, the first thing you have to do is locate a set of solar panel drawings or plans or a kit. You may be able to get these from your hobbyist store too or you can get them from a website that specialises in sustainable or alternative energy sources.

Once you have all your parts and your plans, it will only take you a couple of free hours to put together your solar panel and have it working for you. The next one will take even less time as you get used to it

Renewable Energy

7. How To Save Energy

Our household energy bill, the amalgamation of the electric and gas bills, is by far the biggest bill in our lives. The mortgage may cost more, but at least you end up with a building, the money spent paying the combined energy bill just goes up in smoke.

However, could you envisage a life without energy? It would mean going back a hundred years to when the average household had neither gas nor electricity. If you can cut down on your energy consumption and thus your energy bill, it could create some terrific savings.

Just about half of our energy bill is made up of using our heating and cooling equipment. Therefore, this is the area to start making your savings. The first thing to do is make sure you get value for your money by enabling the heating and cooling systems to give value for money.

Therefore, clean your blower's filters at least once a month, so that it does not have to work so hard blowing air through the filter. Check your radiators or grilles at least twice a year too.

Make sure that they are not heavy with dust or even blocked. Grilles should be meticulously cleaned and vacuumed. Radiators should be washed and drained of air. Make sure that the heat from your radiators is not going up behind the curtain just to keep the window warm. Do not stand furnishings in front of radiators or over under floor heating grilles.

Inspect the settings on your thermostats. Aim to make do with one degree less of heating and let your room warm up one degree when cooling. I assure you, you will not notice the change on your skin, but you will in your pocket. Wear a cardigan in the winter and a thinner shirt in the summer.

If you use ventilation fans in the bathroom and kitchen, do not leave them running without cause. Twenty minutes after you have finished cooking or bathing is more than enough.

If you are still using incandescent light bulbs, change

them for long-life, low-energy fluorescent tubes. Turn incandescents off when you leave the room, but fluorescent tubes cost more to turn on than they do to leave on, within reason. This suggestion can save you a lot of money every year.

Work by a window, if you can. Draw the curtains fully and draw the nets too in order to get the greatest amount of energy-saving, free daylight.

Turn appliances off at the mains and unplug them when not in use. Stand-by uses more electricity than just keeping that little red light on, a great deal more. The same with battery chargers. Phone battery chargers use up energy even when there is no battery in the charger.

Doing the washing is an area for significant savings, especially if you use the washing machine every day. Use a cold water powder and you will save a fortune on heating up the water - as much as 90% of the washing costs. Always wash with full loads or decrease the amount of water to suit the amount of washing.

When you go to the fridge, close the door right away. Do not hold it open while you are drinking or talking, this goes double for the freezer. Check the doors' seals for leaks or cracks. If ants have got in, air has got out. Try to keep your fridge/freezer full; it works out cheaper than keeping litres and litres of air cold.

Insulate your house appropriately. Insulate the doors and windows to stop draughts, but most of all, insulate the loft. It produces the biggest bang per buck in household energy saving. If you have a basement or a cavity floor, take care to plug draughts in that too by placing newspapers under your carpets.

8. Nuclear Energy: To Be Or Not To Be?

There is no doubt that we have to stop burning fossil fuels, not only because of global warming but because the fossil fuels are actually running out and becoming too expensive. The question is: what are we going to use instead? Solar, sea and wind power are all very tempting but at the moment they are probably not reliable enough to base a national economy on.

In this case, the heavy lifting with regard to providing electricity for the national grid would probably have to be done by modern nuclear power plants. However, that does not mean that localities should not use all the resources at their disposal. Quite the contrary, local communities should seek ways of getting off the grid by generating electricity from more traditional alternative energy sources.

The main problems of nuclear power plants as far as

its detractors are concerned are the chances of a catastrophe such as happened in the Ukraine and Japan and what to do with the waste. These two items were / are scary enough to put nuclear power out of the game, but public opinion seems to be turning again.

France has a very successful nuclear programme and so does Finland and people are starting to wonder whether some countries have enough space to be covered with all the wind farms and solar panels that they would need to run a modern economy that is hooked on energy consumption.

Perhaps atomic energy generated by new nuclear power plants can be part of the solution after all. Nuclear reactor technology has advanced a great deal since the early days of atomic energy.

Nuclear radioactive waste is still a major problem that has to be overcome. Up until now, and for the foreseeable future, governments are just encapsulating the waste to make it very safe in the hope that in the future some method will be discovered of rendering the stuff totally harmless. That is their 'policy'.

Surely future scientists will come up with a solution, but is it morally justifiable to produce such toxic material and leave it to future generations to sort out for us? Surely they will have their own problems to deal with and do we really want to be one of them?

Countries have tried burying their radioactive waste in mountains, dropping it in the sea and even selling it to poor countries, but these are not real solutions. The fact is that no-one wants to live near a nuclear waste fuel dump because no-one trusts what the government tells us any more They have been caught out lying too often on far less important issues.

Unfortunately, we have moved too late to have many options left to us. Fossil fuels are running out fast; the governments, that control where most of those fuels are, are unstable and could soon become hostile to the West; the price of energy is so high that many people are suffering and the global environment is already badly damaged from our pollution causing climate changes.

Thirty years ago was the time to move on this, now we are way late and we need to take action right now. It looks as if building nuclear powers stations will be part of the programme - it will have to be - and that

we will be cursed by distant generations for having
left them our mess to clean up.

9. Harnessing The Power Of The Sun To Make Solar Electricity

Free energy ... What a dream, eh? One of the biggest household burdens is the cost of energy. The cost of energy is frequently 40% of total domestic bills. So, free energy would assist every family that is not rich a great deal. However, free energy is a pipe dream, is it not? There is alternative energy, that is non fossil fuel based energy, like nuclear energy, but that is not cheap either.

Other alternative sources of energy include wind-driven turbines and solar power. In this piece, I want to talk about harnessing the power of the sun to make solar electricity. Creating solar power is nothing new and most people are acquainted with the general theory of how the system works. In deed, most of us have owned a solar powered pocket calculator or solar powered clock at one time or another.

Renewable Energy

Solar electricity is just as good and just as strong as conventionally generated electricity and they can be used for exactly the same purposes. However, solar energy has one huge advantage; it is not 'dirty'.

Electricity created from the sun's energy has not been made creating any pollution whatsoever. Furthermore, because there are no moving parts in a solar panel, there is no wear and tear and so less repairs.

Solar panel systems are more adaptable too. For instance, if you have a small home with couple of appliances, you still have to have the same method of delivering grid electricity as a huge house and you still have to have a metering system and a means of paying for the electricity used.

However, if you take the same small house as an example, you might find that ten solar panels will run it. Therefore, for a one-off payment, you are free of electricity pylons and their cables, the meter box and the monthly bills. A huge house would just have to fit more panels, say a hundred, to achieve the same freedom.

This freedom from the means of delivering electricity is a very real advantage if you live in a remote place, where you are expected to pay for the electricity pylons and their cables all on your own. The down side of using solar power is the cost of setting it up. A professionally installed solar energy system can cost about $30,000.

If you save $200 per month on electricity, then you will recoup your expenditure in about 300 months, which is 12.5 years. However, if you could get the system installed more economically, you would recoup your costs more quickly.

This is possible, by assembling the solar panels yourself and installing them yourself. No matter what sort of a ham-fisted person you think you are, you can assemble and install the average solar panel kit. In deed, most teenagers can manage the job.

If you decide to buy solar panel kits to assemble yourself, you can save about half of the above outlay, but if you were to make the panels from parts that are easily obtainable in DIY shops, you could be harnessing the power of the sun to make solar energy for up to 75% of the cost of a professional installation.

Renewable Energy

10. Building Your Own Solar Panels

Everybody these days is aware of the problems of climate change, or as the boffins used to call it, global warming. It seems that even those people who call themselves experts do not know what is going on. It is not very heartening. However, what everyone does know, even without so-called expert advice, is that the rate that we are using fossil fuel up is untenable, that that means that prices will go up and that it looks like the West is on the slippery slope.

So, what can we do about it? People are looking for guidance and help to reduce their dependency on conventional sources of energy like the national power grid and the petrol station, but these people are loathe to give any real advice lest they lose their hold on society.

People are beginning to want to make themselves less reliant on the traditional power suppliers and would like to go it alone. Regrettably, most people do not know where to turn for advice, but they have some imprecise idea about putting up their own solar panels.

However, without any easily understandable government subsidy or recommendations, people are left to do their best alone. And the fact is that the professional installation of your own solar panels is expensive. It costs thousands of Dollars, Euros or Pounds, wherever you are, which is why some people are looking into a do-it-yourself (DIY) solar energy system.

Unfortunately, without any appropriate guidance, home-made solar panels will barely run a low energy light bulb in the midday sun. A better way is to purchase a self-assembly kit from a decent company and put it together yourself, following their instructions. After all, we all know that the labour constituent of any job is often equal to or more than the cost of the parts being fitted. These kits come with everything you need to install your own solar panels.

These self assembly solar power kits are well put together and the instructions are typically well written and easy to follow. However, it is worth buying from one of the better known, big firms like GE, but you can do this from a website that specialises in home sustainable energy systems.

You may like to turn the assembly of a few of your own solar power units into a family venture, because that will teach the younger members of the family about how worthwhile energy conservation is and introduce them to electronics.

These home solar panel systems are completely expandable, so you do not have to deprive yourself of that new dishwasher, although the fact is that modern appliances are using less energy than their older counterparts did.

Just picture if you had a bank of your own solar panels and you became 50% independent of the national grid. And then you added a few more panels when you had the time and in due course became totally independent of the grid! And then you began adding a few more and were in point of fact selling electricity back into the grid. It is all possible when you have your own solar panels.

Renewable Energy

11. *Is A Solar Panel Electrical System Right For You?*

Until approximately a hundred years ago in the West, people only had recourse to renewable energy for heat and light for their homes. They burnt wood and sometimes coal or peat (OK, fossil fuels) and got up when the sun came up and went to bed with the sun as well. In, fact a large part of the world's population still lives like that.

Things changed with mechanised industry and night shifts. Electricity providers sold the populace on being able to do more instead of just sleeping when it got dark, and the Western population got hooked on buying huge amounts of energy, mostly electricity and engine fuel, which was usually produced from oil and coal.

This idea soon travelled around the world and with rising prosperity came emulation and other countries wanted the same. Now we are in the sad situation where we have to confess that we rode the fossil fuel gravy train to its last stop without thinking about what we would use when fossil fuels ran out.

This is where the typical citizen comes in. You have to think about how you want to draw energy in the future. Do you want to be powered by keeping sucking non-renewable resources out of the planet, or do you want to have as little to do with it as you can? Would you rather have everything you have now, but know that the resources that are powering your lifestyle are renewable?

If, like millions of others around the world, you would rather say 'No!' to traditional power production methods, then you have to take a stand. But not only in words, you really have to do some something about it physically.

This will mean paying a lot of money up front, which may not be a problem for you or you may even think that taking a stand is worth looking for a bank loan. These are commendable feelings, but I would like to suggest that there is another way to self-sufficiency.

You could make your own!

Why not? The technology has been around for decades and is fairly easy. Most reasonably competent teenagers can put together a bank of photovoltaic cells into a solar panel and then plug that into your home's electrical system. And if a teenager can manage it, so can you. All you (and the teenager) will require is a solar panel kit and a schematic diagram. A plan in other words.

A solar panel kit can be bought locally from a DIY shop or from the Internet. A typical solar panel will take a few hours to assemble and will produce 100 watts of electrical energy. The electricity produced from these panels is then passed through an inverter that changes the current from DC to AC, making it usable by household appliances and the utility grid.

Do yourself and the planet a good turn, get off the grid and start saving money and the planet's resources, you will be surprised how easy it is once you get started. And do not forget, you can do it in stages of, say, one 100 watt panel a month until you hit self-sufficiency. It is not a question of 'All or Nothing'.

Renewable Energy

12. It Is Time To Utilise Renewable Solar Energy

The very life of every living thing on Earth is reliant on the Sun. Without the Sun, we would not have animal or plant life, which provides us with food and companionship. Without the Sun, we would have died out in earlier days, without the Sun these days, we may last a while, but it would not be much of a life.

The Sun gives us much more too. What we now call 'alternative, sustainable energy' was the only type of energy available to the world for thousands of years. Really, nuclear and even oil generated energy should be called alternative, but we have forgotten a lot. Particularly in the West.

It is not necessary to go back to the habits of our farming ancestors and get up and go to bed with the Sun, although it is still the way of life of the majority of the world's inhabitants. No, we have technology and we should use it. At the moment, we use sophisticated technology to find more lakes of oil and dig ever deeper in more and more inaccessible places to remove it from the Earth. Or worse still, we go to war to steal or ensure supplies causing the death of thousands of young soldiers and the misery of millions of ordinary, innocent citizens.

We need to use our fantastic advances in technology to produce electricity out of thin air. Literally. We already have the technology to produce solar panels and wind turbines in order to create millions of kilowatts of electricity from the Sun and the wind. There is other technology that can make use of the movement of the seas and the natural heat of the Earth itself, although some of these are only accessible in some areas. For instance, wave power can only be taken advantage of, if you live on the shoreline.

However, solar and wind power can be used in any part of the world with varying levels of success. A combination of the two varieties of power generators is probably best for most areas. These technologies have been developed more or less chaotically. What if we had spent our war chests of billions of dollars on progressing these technologies, instead of using them to flatten cities and kill people?

However, the technology is there to generate enough electricity to run a household. It is obvious that we cannot wait for our governments to do much more for us. Their viewpoint is not the same; they do not want to damage big business. And the big electrical providers would be harmed if a substantial number of people generated their own electricity and came off the grid.

The rich representatives in government are stuck between a rock and a very hard place. They recognise that oil is running out; they know that much more electricity will have to be produced from the wind and the Sun, but they do not want to harm the share value of big industry.

Picture, what a hard time the energy suppliers would have explaining why the cost of electricity had to rise because he Sun or the wind had increased its charges. Whereas it is simple to explain when they say that OPEC has raised the price of oil. But, we are in OPEC, aren't we?

Renewable Energy

13. Solar-Powered Lighting

Mankind desires light. We humans are reliant on light unlike some other species. Our whole world existed only in the day light for millions of years. We woke up with the sunlight and went to sleep when the sun went down. Afterward, we learned to produce our own light, but we were not totally successful at it until we started using electricity to produce light.

This has worked well for almost a hundred years, but now we know that the main fuel used to produce our electricity, that is to say oil, is becoming too complicated to obtain and more and more wars are being fought to ensure the West's supply of oil.

Not only that, but greater and greater risks are being taken by oil companies, which has lead to more and more catastrophes like that of the deep sea rig off Louisiana in the USA.

Therefore, the time has come, some even say that it is long past due, that we rely on a different source of power and less on oil. The easiest and least risky way to produce electricity at the moment seems to be solar power. The technology has been with us for decades, but it was not practical for households to use it because it was too expensive, but that is no longer true.

The capability to use solar power to produce electricity for our lighting and indeed all our requirements is there, already on hand. The only difficulty is that there are many oil baron billionaires who do not see this advancement as being in their interest. And they are right, in a way, it is not in their selfish interest to decrease our dependence on oil, but it is in the planet's and its population's interest. That is mankind's predicament at the moment.

Solar panels can be utilised to produce electricity for immediate use either inside or outside the home and the surplus generated can either be stored in batteries or fed back into the electric grid for which you will be remunerated (and compensated at a very high rate, in some countries).

If you do not want to come off the grid just quite yet, you could still use solar panels to run some of your external lights. In deed, there is a broad choice of outdoor lighting that provides its own power.

These external lights use mini solar panels, which the manufacturer has built in to the casing of the light to create power from the sunlight during daylight hours and then stores it in internal batteries to be used later.

These lights also have a capacity to run the light on battery power from dusk until dawn. The solar unit determines whether to charge the batteries or supply light by the strength of the ambient light. This solar power set up is ideal for external security lighting and accent lighting in the garden. They are perfect for lighting up the fish pond or the drive too.

Solar powered outdoor lighting has come a long way and it is well worth you finding out more, if you would like to decrease your dependence on electricity produced by oil from the grid.

Renewable Energy

14. Saving Energy - Saving Money

Energy prices will increase in the long term, we all realise that. It is only temporary relief, when the price of a barrel of oil falls from $150 to $75, we all know that it will go back there. Besides human greed pushing the price up, there are more than three billion people in Asia all wanting to improve their lifestyles to what they see the West parading in its films and TV soaps. And that is not even including Africa and South America. No matter what oil is left undiscovered under the soil of the globe, it is not enough.

So, what can you do about it? Use less, is one obvious answer, but it is hard to give up something you were born into or have become accustomed to over a protracted period of time. It is just not that easy. We can expect to see equipment that will use less energy than they do now. That will help, but the technology is still being developed. The only option left is to be far more careful with the energy at our disposal. Turning lights out is the most simple form of this way of saving energy.

In the long term, the government will have to set standards for manufacturers, including, and actually, especially for house builders. Nearly 50% of household fuel budgets go on heating and cooling. Solar panels built into the roof would help a great deal, but they are still too expensive for most people at the moment. These items have to be made affordable to homeowners and they have to be incorporated into all new homes.

Solar panels can be utilised to run devices live, to put power back into the electricity grid, if their is no immediate local need for it or to recharge batteries to run equipment later, such as low power lights, a hybrid car or an electric scooter. This would mean a huge personal and national energy saving, but the sun's energy can do more than that.

Solar heaters can be used to heat both air and water. This would chiefly eliminate the need for burning fossil fuels and natural gas. I say largely eliminate not eliminate, but it would be a massive saving again. Do not forget that nearly 50% of the household fuel budget goes on heating and cooling and with 'global warming' or 'global cooling' or the very safe, sit on the fence 'global climate change', this percentage can only rise.

Geo-thermal power (getting heat out of the ground) is probably not an alternative for every area, but it is a source of power that has hardly been touched in most countries. Greenland and Australia are world leaders in this technology, I believe, and their climates seen to be as much at opposite ends of the scale as their locations are at opposite ends of the planet. There must be more that could be done with this technology.

In the meantime, we have to do what we can. We have to re-educate ourselves to be more careful with energy and we have to teach our young from an early age to be careful with it as well. Fit energy-efficient, fluorescent tubes wherever you can. If you need more light for reading or working, get a desk lamp. Turn things off when they are not in use. Stand-by was a great idea, but it is not any more.

Items on stand-by are still drawing electricity and not just a tiny bit to keep that red light on. Chargers for mobile phones et cetera draw power even when there is nothing plugged into them to charged. Take them out of the socket when not in use. Integrate your heating and cooling systems. There is a lot you can do to save energy and save money, if you put your mind to it.

15. The Best Way Get Off The Grid

I don't know whether you have ever thought of it, but you can get off the grid. The electricity grid that is. Is it anything that you are interested in? For some people, getting off the grid is the Holy Grail of modern life. We all believe that we are being milked dry by the energy companies and our soldiers are being killed on a daily basis to secure the West's supply of oil.

If the same amount of money was invested in developing alternative technology, we would not have to import so much oil, but that would mean less profits for the energy bosses too. So the average guy on the street loses out to the rich oil tycoons and people are squeezed for money and people die on both sides.

You cannot rely on these people to help you use less energy, it is not in their interest. The only thing you can do is take the bull by the horns and sort yourself out. Get yourself off the grid, save yourself money in the long term and help save the planet and our soldiers too.

It is often assumed that self-sufficiency in energy is only for geeks, but this is not true, it is now within the bounds of every family or household. If you have plenty of money, a solar or wind powered system will cost you about $45,000 and it will take about ten years to recoup your investment by getting off the grid. However, you can have the same apparatus for much less than half that sum.

The fact is that technology has advanced rapidly in the field of solar power, as it has in every other technological sphere. Scorching sunshine is no longer essential to create a steady flow of electricity and prices have dropped. However, labour prices have risen and over-compensated for this fall. Herein lies your opportunity.

It is not hard to make your own solar panels and the components are readily obtainable in even small towns. The same holds true for systems that will create electricity from wind or water. The trick is to get a detailed diagram from which to build your solar panels or other equipment.

These designs are readily available online from specialist web sites and the components are available from DIY or hobbyist stores. Once you have made your sustainable energy resources, they are practically maintenance free, although you may have to top up the batteries with sulphuric acid or distilled water.

Living off the grid is a magnificent feeling. Knowing that you are out of the grasp of the oil barons brings more freedom and the next time there is a power cut or a price hike, you can just give a smile and forget about it, because you will be living off the grid.

Renewable Energy

16. Wind Turbines - Why Not Build Your Own?

What is your first impression of wind turbines? Do you think that the people who make them must be highly qualified technicians who have studied for years?

What would you say, if one of your neighbours said that they were going to assemble their own wind turbines in order to produce their own electricity?

The fact is that it is not really very hard to make a wind turbine. You certainly do not have to have been to university to do it and you do not have to be an amateur electrician either. You will, however, need a good set of plans in order to put up an effective wind turbine.

There are various sources of plans for wind turbines. You could go to a hobbyist shop, a DIY warehouse or the Internet. If you search on the Internet, search for a web site that specialises in alternative energy technology.

Once you have you manual, guide or drawings for making wind turbines, you should study it well and get any questions you have answered on forums on the Internet.

If you believe that I am exaggerating how simple it is to build a wind turbine, do not forget that people have been building windmills and water pumps for hundreds of years without the advantage of power tools or modern materials.

The only dissimilarity between those early devices and a wind turbine for producing electricity is the addition of wiring and a coil, which you will buy anyway.

The basics of building a wind turbine involve building a tower to place it on; installing batteries to take the generated power; the fan and the tail assemblies. The tools that you will require to accomplish this are fairly basic too: bricklaying tools, carpentry tools and a few spanners, but that really depends on how much you sub-contract out.

You will almost surely have to buy the turbine, fan and tail section yourself as one element, because, if you want a high-powered unit, this section will have to be professional, not that it cannot be manufactured locally.

You will be able to maintain it yourself as well, if you want. These fan blades can be 1.2 metres (4 feet) in length each, which is something that you will have to keep in mind when building your tower.

You may require a bit of professional assistance when you build your first wind turbine, but use the experience to learn as much as you can, so that you will be able to be more independent on your next one. You could even turn the experience into a profession, because there is a chronic scarcity of people who know anything about wind turbines.

Renewable Energy

17. Learn How To Make Solar Panels

Many people are sick to death of the escalating cost of electricity, but they do not know what they can do about it. They have switched lights off; they have not switched lights on; they have unplugged devices on stand-by and they have turned the central heating down and the air con up and it has only saved them pennies. Yet still the electricity bills go up on a regular basis. It is very exasperating.

We have all heard of alternative and sustainable sources of energy, but the question is: how can average people implement these technologies into their own homes? One phone call to a professional solar power installation company will convince anyone that solar power is only for the well-off.

Similarly with wind power generation. A single device, which may or may not satisfy the electrical requirements of one average family will cost in the region of $45,000 and it is thought that, if it meets aspirations, that it will pay for itself after ten years time. $45,000 is too much money for most individuals, but having to wait ten years to realise that investment is too long for most people too.

So, at first glimpse, it looks like the average person is at the mercy of the oil companies and the electricity generators. However, it does not have to be like that. There are other ways of tackling the problem and extricating yourself from the clutches of energy providers.

The first technique would be to buy the solar power units and install them yourself. That would save you a great deal, perhaps up to 50%. But you could go a stage further and make the solar panels yourself too, which would save you a great deal more money.

The fact is that the more of your own labour that you can put into the scheme, the more you will save, because the labour constituent in any task usually amounts to about 50% of the costs. We all know how costly it is to call a plumber in to fix a leaking pipe or a roofer to replace a slate. $100 for a slate? $100 for a dribble of solder? No, it is the tradesman's time and the same is true of solar panel fitters.

The way out is to learn how to make the solar power units and install them yourself. It is not difficult. Any teenager can learn how to do it and so can you. Take it from me, learning how to assemble and install solar panels is not difficult. Just try it and see for yourself. You will amaze yourself and your friends. You may even turn it into a business!

The thing to do is get a reliable kit or a plan from a trusted source. This could be a neighbourhood hardware or hobbyist store or from a specialist website on the Internet. Once you have your design, you can buy the pieces of equipment. They are commonplace and on hand in most DIY shops. Then it will take you about a day to make a solar panel that will provide about 100 watts.

This may not sound like much, but it will power many kinds of small machines or two average light bulbs. The trick is to keep adding to your bank of solar panels until you are entirely free of the electricity grid or even selling electricity back into it.

Renewable Energy

18. *What To Know Before Purchasing Home Solar Electricity*

The replacement of conventionally created electricity with electricity made from renewable or different resources is now a feasible alternative for many people in the Western world. The major stumbling block is the initial outlay, which is why solar electricity is still not a viable affair in most other, hotter, countries.

The fact is that fitting solar panels to get your home off the grid is a lot less expensive than it was ten years ago, but it is still not cheap. Some countries have introduced encouragement schemes and these are fine, as far as they go, but often they are designed for the middle classes, which is not a segment of society as big as the working class and which can afford to pay for its own electricity anyway. These schemes leave the preponderance of the members of society stuck with the grid. The new British scheme FITS is like this.

Other countries have so-called 'Green Options', which means that you can decide to draw energy from producers of electricity from users of renewable resources only, which is great as far as it goes, but the end user is still trapped in the system of being on the grid and being subject to price hikes and power cuts.

If you really want to get off the grid, do away with monthly bills and recover your freedom from the fat cat oil and electricity suppliers, you need to take a radical approach. The first step is to work out your electrical needs.

Work out the coldest and the hottest month and use the dearest plus 10% as your goal. The fact is that it could take you years to get off the grid, and by then white goods will be using less electricity than they do now anyway. You can also sell your surplus electricity back to the grid for real happiness.

The cost of the professional installation of solar energy systems can be prohibitive and take twelve years or more to recoup, but if you assemble and install your own bank of solar panels, you can more than half that figure. In deed, it is possible to decrease the cost by as much as 75%, if you are willing to assemble and install the solar panels yourself. This is a task that most capable teenagers can do, given the right drawings or schemas.

The best way of going about it, is for you to read up as much as you can on the topic, because there are several routes you can go. The main ones, using solar panels or other techniques of renewable electricity, are: remain linked up to the grid, using your own electricity first and selling back any surplus; remain linked up, but send surplus electricity to your own batteries, which could be an electric car; or you can get off the grid altogether.

The ultimate goal, in my eyes, is to supply my house with all the home-made electricity from solar panels that I require, to recharge my hybrid car's batteries from the same source and to resell any surplus back to the electricity grid.

What a dream!

Renewable Energy

19. Solar Panels For Home Or Business Energy

It is almost high noon on the energy front. Oil, the material that is used to produce most of the world's electricity is becoming too expensive to use to create electricity for home consumption. Even if you think that there are lakes of oil left, which could be true, the main reason they have not been exploited yet is because it is too costly to get out. The only thing that makes it a viable concern is the high price of oil on the market.

Therefore, it stands to reason that oil prices cannot go down in the long term, which means that our electricity and petrol prices will remain high and will almost certainly keep going up. Add to that the fact that manufacturing is moving to the Far East and the fact that immigration is increasing and the result is lower wages in the West. The likelihood is that the average national wage will not rise as quickly as the price of energy.

So what can you do about it? Well, while commodity prices are bound to keep rising, one thing has always kept falling and that is the cost of new technology. Or to be more precise, slightly old technology. Cutting edge technology is always expensive, but after a few years the price falls, as we saw with desktop computers and as we are seeing with laptop computers now.

The same trend is at work with solar panels. They are far less expensive now than they were a few years ago and they are far more responsive too. And did you know that the bits and pieces that are used to make solar panels can be bought from plastic bins at most DIY and hobbyist stores like Radio Shack? If you knew what to buy you could literally go out and bring back enough bits and pieces to make a few solar panels the next time you go out for a loaf of bread.

So why are we not doing it? We did not know that was feasible? Nobody told us? We are not technologically minded? We do not have the skill?

OK, all those reasons sound valid. Nobody has been

telling us, but the truth is that it is straightforward to make solar panels and not that dear any more. Professional installations are still dreadfully expensive - about $45,000 -, but you can do it yourself. There are two approaches you can take.

You can either get a schematic diagram, a plan, from a hobbyist shop or the Internet, purchase the parts and make your panel or you can buy a self-assembly kit. Sincerely, there are kits about that teenagers can assemble as easily as they do a plastic model aeroplane. 'Locate and insert part number 44 into the main board number 3' - that sort of simple.

If you are new to the world of solar energy, then you may be asking yourself how solar energy panels work. Solar energy panels are also known as photovoltaic panels; photovoltaic meaning electricity from light. Solar energy panels work by collecting protons from the sun, which dislodge neutrons, and thereby generate a flow of electrons or electricity. This electricity can either be stored in batteries for later use or used directly.

You can utilise solar panels to heat your pool, run your workshop tools, power the greenhouse lights and fans or if your system is big enough, replace grid electricity in your entire home or business. Most solar energy panels are intended to last upwards of 20 years but involve little to no upkeep.

There is a drop-off of the power supply after about 10 years of about 10%, but over the life-span of the solar panels, the energy savings made are enough to recoup the initial cost of the system and more. Furthermore, costs are tumbling while energy prices are rising. It already makes economic sense to change to solar energy power.

20. Top Energy-saving Tips

In today's world, the imperative to conserve energy and adopt sustainable practices is more pressing than ever. Not only does it contribute to a healthier environment, but it can also lead to significant cost savings. Whether you're a homeowner, a business owner, or simply someone who cares about the planet, implementing energy-saving measures can make a meaningful impact. Here are some top tips to help you on your journey towards a more sustainable and energy-efficient lifestyle:

Upgrade to LED Lighting: Swapping out traditional incandescent bulbs for energy-efficient LED lights is one of the quickest and easiest ways to save energy. LEDs use significantly less electricity and have a longer lifespan, reducing both your energy bills and the need for frequent bulb replacements.

Seal Air Leaks: Gaps and cracks in windows, doors,

and walls can lead to significant energy wastage. Seal these areas with weather stripping or caulking to prevent heated or cooled air from escaping. This simple step can result in substantial energy savings over time.

Invest in Energy-Efficient Appliances: When it's time to replace appliances, opt for those with high Energy Star ratings. These appliances are designed to use less energy while performing the same tasks, ultimately reducing your electricity consumption.

Unplug Electronics: Many electronic devices continue to draw power even when they're turned off. Combat this "phantom energy" by unplugging devices or using power strips with on/off switches to completely cut power when not in use.

Optimise Heating and Cooling Systems: Regular maintenance of your HVAC system, including cleaning or replacing filters, can improve its efficiency. Additionally, consider installing a programmable thermostat to regulate temperatures based on your schedule, avoiding unnecessary energy consumption.

Harness Natural Light: Take advantage of natural

light during the day by opening curtains or blinds. This reduces the need for artificial lighting and creates a more pleasant living or working environment.

Utilise Smart Technology: Smart home devices, such as smart thermostats, lighting systems, and power strips, allow you to monitor and control energy usage remotely. This technology empowers you to make real-time adjustments to optimise energy efficiency.

Opt for Renewable Energy Sources: If possible, consider installing solar panels or utilising other renewable energy sources. This can significantly reduce reliance on fossil fuels and lower your overall carbon footprint.

Upgrade Insulation: Well-insulated homes are more energy-efficient, as they retain heat in winter and keep interiors cooler in summer. Consider adding or upgrading insulation in your walls, attic, and floors to create a more thermally efficient space.

Conserve Water: While not directly related to electricity, water heating is a significant energy expense. Use low-flow fixtures, fix leaks promptly, and consider installing an energy-efficient water heater to reduce water-related energy consumption.

Choose Energy-Efficient Windows: High-performance windows with multiple panes, low-E coatings, and gas fills can significantly improve your home's insulation and reduce energy loss.

Promote Natural Ventilation: During mild weather, take advantage of natural breezes to cool your home or workspace. Strategically positioning windows and using window fans can help circulate fresh air without relying on air conditioning.

By incorporating these energy-saving tips into your daily routine, you can contribute to a more sustainable future while enjoying the financial benefits of reduced energy bills. Remember, every small effort counts, and collectively, we can make a significant impact on energy conservation and environmental preservation.

Owen Jones

Contact Details

Facebook: AngunJones
Twitter: @owen_author
Blog: Megan Publishing Services

This book is part of the 'How to...' series of 150 manuals by Owen Jones.
The whole series can be found in many languages on Megan Publishing Services at:
https://meganthemisconception.com